Growing and Curing Sun-Cured Tobacco

by Virginia Polytechnic Institute Agricultural Station

with an introduction by Roger Chambers

This work contains material that was originally published in 1912.

This publication was created and published for the public benefit, utilizing public funding and is within the Public Domain.

This edition is reprinted for educational purposes and in accordance with all applicable Federal Laws.

Introduction Copyright 2018 by Roger Chambers

IMPORTANT NOTE & DISCLAIMER

IMPORTANT NOTE :

As with all reprinted books of this age that are intended to perfectly reproduce the original edition, considerable pains and effort had to be undertaken to correct fading and sometimes outright damage to existing proofs of this title.

At times, this task can be quite monumental, requiring an almost total rebuilding of some pages from digital proofs of multiple copies. Despite this, imperfections still sometimes exist in the final proof and may detract slightly from the visual appearance of the text.

Some images may suffer from reduced quality due to anomalies in the original scan.

DISCLAIMER :

Due to the age of this book, some methods or practices may have been deemed unsafe or unacceptable in the interim years. In utilizing the information herein, you do so at your own risk.

We republish antiquarian books with no judgment or revisionism, solely for their historical and cultural importance, and for educational purposes.

Self Reliance Books

Get more historic titles on animal and stock breeding, gardening and old fashioned skills by visiting us at:

http://selfreliancebooks.blogspot.com/

Disclaimer

This book was written in an age when little was known about the ill effects of tobacco.

The material presented herein is intended to be strictly for educational purposes with the purpose of enlightening readers about the historical uses of tobacco. Publication of the material is neither an endorsement, nor a criticism of its contents. This book is presented as part of large series of educational material on the history and cultivation of tobacco.

As the reader, please consider it your duty to consult with a medical doctor before utilizing tobacco. It is also the reader's duty to become familiar with local, state, provincial and federal laws relating to the growing of tobacco.

As the author, publisher and retailer cannot control how the reader utilizes the historical information presented in the pages herein, they hereby disclaim any liability to any party for any loss, damage, disruption, death or other liability that may be incurred by the reader's misuse of this material.

introduction

Here at **Self-Reliance Books** we are dedicated to bringing you the best in *dusty-old-book-knowledge* to help you in your quest for self-sufficiency and independence.

We're so pleased to bring you this old title on tobacco production. These old reports and bulletins put out by the USDA, other government departments, and educational institutes are very popular. It should be said, though, that some of the information is best looked at in the historical aspect, due to the obsolescence of some practices or methods.

This special edition of **Growing and Curing Sun-Cured Tobacco** was written by W.W. Green of the *Virginia Polytechnic Institute Agricultural Station*, under the direction of E.H. Mathewson from the *Office of Tobacco Investigations,* part of the *U.S. Department of Agriculture*. It was first published in 1912, making it well over a century old. It is also known as **Bulletin 197**.

This super-short, fast read features sections on *Varieties of Tobacco for the Sun-Cured District, Saving Tobacco Seed, The Seed Bed, Fertilizing Sun-Cured Tobacco, Cultivation of Tobacco*, and more.

Another great Agricultural Station publication that is a must-have for the libraries of all those interested in the historical aspect of the Tobacco Industry.

~ Roger Chambers

State of Jefferson, March 2018

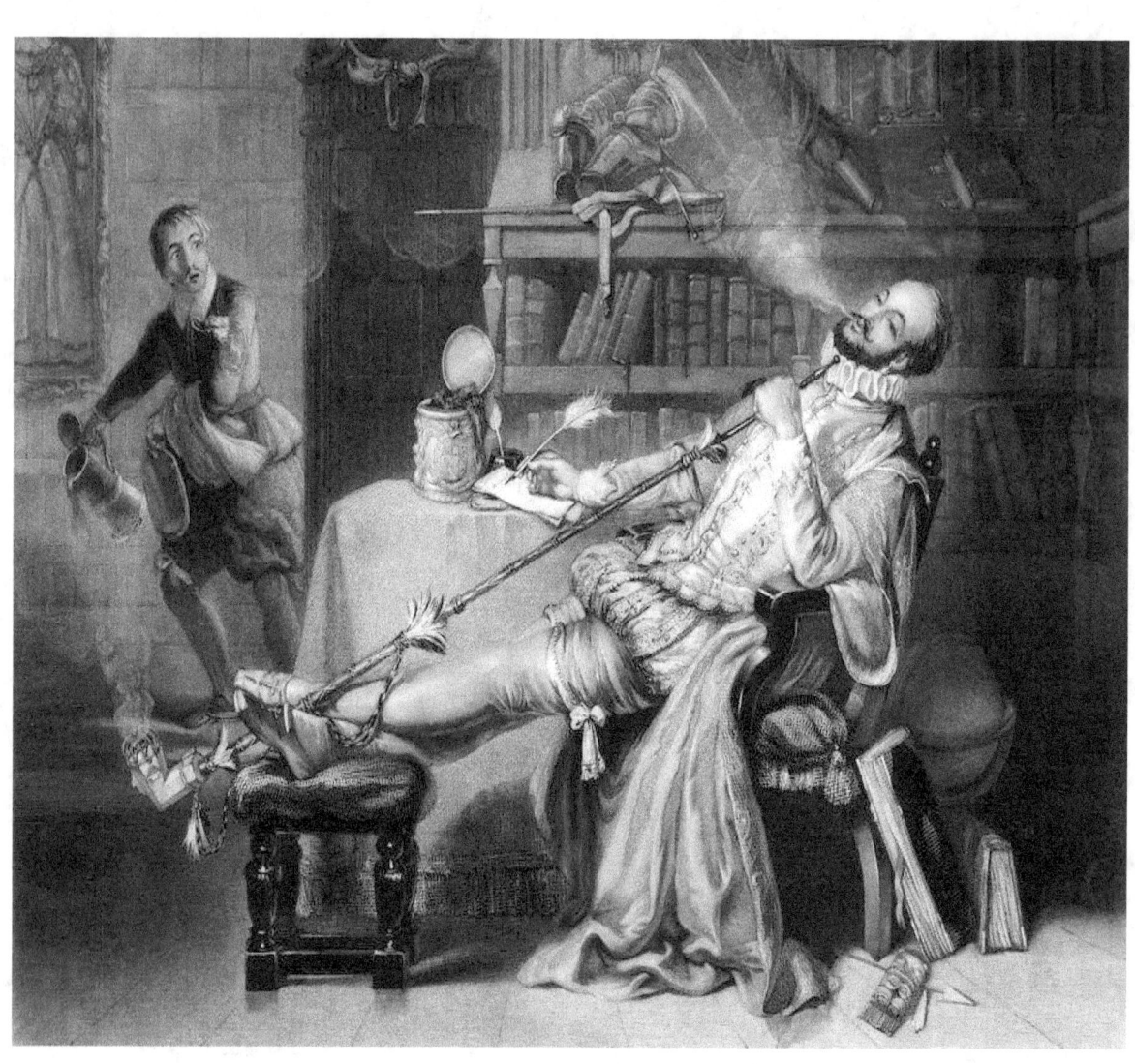

Growing and Curing Sun-Cured Tobacco

By W. W. Green

Under the direction of E. H. Mathewson (representing the Office of Tobacco Investigations, Bureau of Plant Industry, United States Department of Agriculture) and Lyman Carrier (representing the Virginia Agricultural Experiment Station).

The results of certain coöperative experiments in the sun-cured tobacco district of Virginia by the Virginia Agricultural Experiment Station and the Office of Tobacco Investigations, Bureau of Plant Industry, United States Department of Agriculture, are presented in Bulletin 196 of this Station.* The experiment plats are located at Bowling Green, in Caroline County, and at Louisa, in Louisa County. In connection with these experiments there has been opportunity to make a study of the methods of growing and curing sun-cured tobacco. The present Bulletin records the conclusions from this study.

Fig. 1.—Farmers studying the results of the tobacco experiments at Bowling Green. Note the plants bagged for seed.

VARIETIES OF TOBACCO FOR THE SUN-CURED DISTRICT.

In order to determine the comparative merits of the different varieties of tobacco adapted to the sun-cured district, tests of 19 varieties have been made. One hundred plants of each variety were grown. The soil was uniform, and the fertilizer was the same for all. The average value of the crop

---

* "Crop Rotation and Fertilizer Experiments with Sun-cured Tobacco," Bulletin 196, Virginia Experiment Station, by W. W. Green.

of each variety for two years is given below, the figures being corrected to an acre basis. It is admitted that a comparison of so small a number of plants is of doubtful practical value in itself, but by carefully comparing the weights of the varieties and noting the quality for two or three years, four or five of the best varieties can be selected and cultivated on a large enough scale to be sold separately, and thus determine the best varieties.

TABLE I.

Comparative Gross Sales of Varieties of Tobacco.

Varieties tested—	Gross value of crop per acre.	Varieties tested—	Gross value of crop per acre.
Broad Leaf Orinoco	$175.49	Yellow Pryor	$157.22
Narrow Leaf Orinoco	185.41	Kentucky Pryor	189.89
White Stem Orinoco	171.87	Blue Pryor	154.41
Lizard Tail Orinoco	197.16	Warn	174.37
Little Orinoco	147.12	Broad Leaf Tilley	144.56
Hester	158.72	Improved Tilley	144.03
Flannagan	185.15	Smith	172.60
Silver Pryor	151.17	Bonanza	152.29
Hickory Pryor	143.20	Improved Mammoth	165.41
		Conqueror	155.85

This test indicates that the best three varieties are Lizard Tail Orinoco, Kentucky Pryor and Narrow Leaf Orinoco, in the order named. However, this is the result of only two years' tests. These tests should be carried on for a number of years more in order to give thoroughly dependable results. These three, as well as several others on the list, are known to be good varieties for this section, but we would especially recommend the Narrow Leaf Orinoco, as it is extensively planted and seems to have been carefully bred by selection, and usually gives good results. Other varieties have some features superior to the Narrow Leaf Orinoco. The Little Orinoco, on account of its very small stalk and stem, is less susceptible to pole-sweat; while Kentucky Pryor is hardier during its early growth, and will recover better from hardships, such as the effect of grass in wet weather, or lack of cultivation.

SAVING TOBACCO SEED.

At Bowling Green the Narrow Leaf Orinoco was selected as the variety for general planting on the fertilizer plats and on the demonstration fields, as it seemed to be the most popular variety in that section. Each season since special care has been exercised in selecting the seed plants of this variety, the seed being grown under bags. A strain of tobacco of marked uniformity has thus been obtained, every plant being practically identical throughout the entire field where it is planted. This has attracted the attention of farmers, and over 100 tobacco growers in the sun-cured district have requested seed from the Station during each of the last two years, and some of the most prominent seedsmen of the country have written for seed; all of which shows the importance of care in saving seed.

FIG. 2.—Three distinct types of tobacco plants found in a field of tobacco grown from seed produced by one plant, but not protected from crossing with inferior plants; showing the necessity of producing tobacco seed under bags.

In selecting seed plants the first requisite is to fix firmly in mind the qualities desired in the ideal plant. The selection should also be made from the earliest plants, as earliness in maturing is always desirable. When the plant is ready to blossom the seed head should be covered with a 10-pound manila paper bag. This can be done best by drawing the top leaves together and upward with one hand, and placing the bag over the head with the other. Then gather the bag together at the bottom and tie securely. For the first week or two the bag should be raised every second day in order to prevent crowding by the rapid growth. Later the plant should be examined twice a week in order to remove the decaying blossoms, which would cause mould. Bud worms should also be looked for and removed, as they soon bore into the seed pods and destroy them.

The tobacco crop can be materially improved in this way, as it is naturally a self-fertilizing plant, but it can be easily crossed by birds or flies with any variety that may be planted in the neighborhood. Any sucker or poor tobacco plant that may be allowed to bloom at the same time will cross with it if the blossoms are not protected under bags.

Another important point is to clean light and imperfect grains from the seed. This is as necessary as the cleaning of seed wheat. This will be done free for any tobacco grower in the sun-cured district. Address "Experiment Station, Bowling Green, Virginia," and see that the return address is plainly written on the package.

THE SEED BED.

Nothing is more important in the growing of tobacco than to get an early and even stand in the field; this makes the care of the plant bed a very important point. While other methods often succeed, the surest method is to burn well and fertilize well; 100 pounds per 100 square yards of a good "3-8-3" fertilizer will answer. The fertilizer should be thoroughly worked into the soil when preparing the bed, after which it is well to top-dress the bed with 25 pounds of cotton-seed meal per 100 square yards, or its equivalent of nitrogen from some other source.

The bed should be covered with a thick canvas about the 10th to 15th of March, before the seeds have begun to germinate. Many beds that are prepared well otherwise are covered with a piece of old canvas, full of holes, by sticking a peg through it into the ground to hold it down around the edges. This gives the fly free access to the bed. Time and money can not be more profitably spent than in placing 1- x 6-inch boards around the bed on edge, staking them down securely and banking dirt against them to stop all holes. Then cover the bed with thick canvas by rolling a lath or tobacco stick in the edge of the canvas and securing to the board with nails.

FERTILIZING SUN-CURED TOBACCO.

Most farmers wish to get at least 1,600 pounds of a good quality of tobacco per acre. To produce this crop the plant must draw from the soil about 82 pounds of nitrogen,* 12 pounds phosphoric acid and 100 pounds potash. If

---

*The various fertilizing materials that supply the plant food nitrogen do not always furnish it in the form of ammonia, which is only fourteen-seventeenths nitrogen; hence farmers should accustom themselves to using the term nitrogen instead of ammonia. Some fertilizer dealers use the term ammonia, instead of nitrogen, because it makes a larger percentage on the analysis tag and sounds bigger than it really is; it is used to fool the farmer. Farmers should cease to talk about ammonia, and should insist that the fertilizers they buy show the per cent of actual nitrogen. To convert a certain per cent of ammonia into its equivalent of nitrogen, multiply it by .824. To convert a per cent of nitrogen into ammonia, multiply it by 1.214.

this yield is attained the necessary plant food must already be in the soil or it must be put there by an application of fertilizer. While an analysis of the soil may show enough of these elements present, they may not be in a form which the plant can use. Our experiments show that it is necessary and profitable to apply to the tobacco crop about as much nitrogen and potash as the desired crop will need. It is best to apply an excessive amount of phosphoric

FIG. 3.—An acre of sun-cured tobacco on the Bowling Green experiment plats, fertilized with 1,000 pounds cotton-seed meal, 200 pounds nitrate of soda, 600 pounds acid phosphate, and 200 pounds sulphate of potash per acre. Yield, 1,910 pounds per acre. Value, less cost of fertilizer, $200.39.

acid, as compared with the quantity actually assimilated by the crop; only when a large quantity is applied are the best results obtained. Our experiments indicate that it pays to apply as much as 112 pounds of phosphoric acid to the acre for tobacco (furnished in 700 pounds of 16% acid phosphate), which really is about six times as much as the crop uses in producing 2,000 pounds of leaf per acre. What is not used by the crop remains in the soil for succeeding crops.

Fertilizer Recommended.—When it is desired to produce a ton of tobacco to the acre in the sun-cured district (and we have obtained this several times under favorable seasons, once 2,150 pounds, and once 2,260 pounds) we have found it necessary to apply 92 pounds of nitrogen, 112 pounds of phosphoric acid and 100 pounds of potash. This is supplied in the formula recommended in our rotation, as follows:

```
Cotton-seed meal..................................1,000 lbs.
Nitrate of soda ..................................  200 lbs.
Acid phosphate ..................................  600 lbs.
Sulphate of potash ..............................  200 lbs.
```

This is applied at the rate of 2,000 pounds per acre. The analysis of this fertilizer is: Nitrogen, 4.5% (=5.5% ammonia); phosphoric acid, 5.75%; potash, 5%.

This has always given us good profits, after counting all labor at $1.00 per day, single team $1.00, double team $2.00, and the investment in land, taxes, etc., at a fair valuation.

Advantages of Heavy Fertilizing of Tobacco.—The average application of fertilizer for tobacco in the sun-cured district is about 400 pounds per acre of a fertilizer analyzing 2.4% nitrogen (3% ammonia), 8% phosphoric acid and 3% potash, which supplies 9.8 pounds of nitrogen, 32 pounds of phosphoric acid and 12 pounds of potash per acre. There is only enough plant food in this application to produce about 250 pounds of tobacco per acre, and even though a 1,000-pound crop is raised the land is robbed of three-quarters of the plant food that it took to produce it, and is not in shape to produce the other crops in the rotation. Since there is no other crop of high enough money value to justify a heavy application of fertilizer, the land remains poor and eventually may be turned out to gully and grow up in pines for nature to recuperate it.

Below is a comparison of two applications of fertilizer on uniform soil, and under identically the same cultural conditions. One application is the ordinary "3-8-3" fertilizer, but used in even a larger quantity than is customary. The other is mixed to meet the requirements of the tobacco crop. These results were obtained in 1910 on land that would not produce more than seven bushels of corn per acre in a normal season.

Table II.

Comparative Results From Heavy and Light Applications of Fertilizer to Tobacco.

Fertilizer used per acre.	Cost per acre.	Yield per acre.	Gross value per acre.	Value of crop per acre, less cost of Fertilizer.
Cotton-seed meal ..1,500 lbs. Acid phosphate ... 500 lbs. Sulphate potash .. 200 lbs.	$32.41	1,110 lbs.	$112.33	$79.92
Ready-mixed "3-8-3" ....1,000 lbs.	12.50	580 lbs.	51.19	38.69

This shows more than twice the net returns from the heavy application of home-mixed fertilizer than from the "3-8-3," and there was also a marked benefit to the following wheat crop. Where else can the farmer make an investment that will pay 100% in eight months?

Mixed fertilizer costs the farmer from $3.00 to $6.00 per ton more than he can prepare the same plant food for at home. Aside from this there is usually no objection to the mixed fertilizers now on the market. They are almost without exception what they claim to be, and the plant food constituents are usually in an available form. If mixed fertilizers are bought it is important to get one with an analysis that meets the requirements of the crop for which it is intended. To meet the requirements of a 2,000-pound crop of sun-cured tobacco in respect to nitrogen and potash it would take about two tons of a "3-8-3" mixture, which would cost about $50.0′ and there would be three times as much phosphoric acid as is necessary. This would injure the quality of the crop by causing a quick, thin growth, that would fire before ripening. A mixture analyzing 4.1% nitrogen (5% ammonia), 6% phosphoric acid and 8% potash, used at the rate of 1,500 or 2,000 pounds per acre, will give excellent results on tobacco. Fertilizers with this analysis can be found on the market.

Never buy a low grade fertilizer; that is, one with but a small per cent. of plant food; you pay freight on and handle a "filler" that is worthless. When fertilizer is applied to thin land in small quantities for tobacco or corn it is best to put it in a drill or in the furrow before planting; but when large quantities are used, especially on improved land, it is best to apply at least one-half of the amount broadcast. In many cases it is well to apply 100 or 200 pounds of nitrate of soda and 200 to 300 pounds of acid phosphate in the drill, and the rest of the application broadcast.

CULTIVATION OF TOBACCO.

The best results are obtained with tobacco when the crop grows off promptly, with no subsequent checking of the growth. A careful hand-hoeing as soon as the plants take root—a week or ten days after transplanting—will encourage a quick start. The crop should be carefully cultivated after showers as soon as the soil is dry enough to work freely; not so deeply as to break the roots, but thoroughly, about two inches deep, in order to destroy all grass and keep the soil well mulched to prevent the escape of moisture. We prefer the Planet, Jr., cultivator, using the narrow teeth or bits the first few times. Later the wing or mould board can be used, throwing a shallow ridge to the plant. The 18-inch sweep should be used on the rear arm to fill the trenches made by the teeth or mould boards.

Throwing dirt to the row of tobacco with a turning plow should be abandoned, except for an exceptionally cold, damp soil, which is not usually adapted to tobacco, anyhow. This method breaks the roots and exposes too much soil to be dried out by the sun and air, especially in case of a drought.

Cultivation should continue until the crop is well under way, even after

FIG. 4.—The Planet, Jr., type of cultivator, suitable for cultivating tobacco.

the leaves have met across the rows and a large per cent. of the plants have been topped. This can be done without injury by rubbing a little oil or grease of some kind on the horses' legs; also on the traces and whiffle trees. The latter should not be more than fourteen inches long.

TOPPING.

There is no rule as to the number of leaves that should be left on a plant when it is topped. This must be decided by each grower, according to the quantity of fertilizer used and the condition of the soil. Land that would produce five barrels of corn per acre in a normal season without fertilizer might be expected to produce about 900 pounds of tobacco per acre, using the ordinary quantity of fertilizer. On such land about 5,000 plants should be set per acre, and the plants should be topped to about 10 leaves each, giving a total of 50,000 leaves per acre.

Should it be desired to double this yield and get 1,800 pounds of tobacco per acre on the same piece of land, and sufficient fertilizer is applied to do so, the increase in yield should be secured not only by growing the same number of leaves larger and thicker, but also by increasing the number of plants per acre, and the number of leaves per plant. As many as 6,500 plants should be set per acre, and the individual plants should be topped to about 15 or 16 leaves each. In this way the number of leaves to the acre will be increased to 100,000, and the size of the leaves will be held within the limit necessary for good quality in meeting the requirements of the market. In order to double the yield of 900 pounds the fertilizer must be more

Fig. 5.—Method of scaffolding tobacco in the field.

than doubled, and the soil must be well supplied with humus. No more plant food can be expected from the soil itself in producing the 1,800-pound crop than the 900-pound crop, and the moisture supply will remain the same. Experiments have shown that in order to increase the yield to 1,800 pounds the fertilizer would have to be increased, probably, about five times over the amount necessary to produce the 900-pound crop, even though the physical condition of the soil and the moisture supply were sufficient to render this increased yield possible.

FIGHTING THE HORN WORM.

There are usually two broods of horn worms, the first appearing from the middle to the last of June, and the second from five to seven weeks later. The first can be successfully destroyed by spraying with a spray pump, 3 teaspoonsful of Paris green to 5 gallons of water; or by an application of 1 to 1½ pounds of dry Paris green per acre, applied with a powder gun. If applied in water the mixture must be kept well stirred to avoid settling.

When using Paris green to destroy the second brood of worms there is greater danger of burning the leaves. It should not be applied for at least three days after the suckers have been broken off. A safe method of destroying the second brood of worms is to use dry, powdered arsenate of lead at the rate of about 4½ pounds per acre. There is no danger of burning the leaves when this poison is used. It should be mixed with an equal quantity, by weight, of dry, sifted wood ashes as a carrier. The mixture can then be applied best with the dust gun made by Leggett & Bro., New York.

CURING SUN-CURED TOBACCO.

Curing tobacco is a delicate operation requiring great skill. It is not merely a drying process, as is often thought, but a process in which there must be certain chemical changes in order to get the desired color and texture. These changes take place most freely at a temperature ranging from 80° to 90° F., and in a humidity of about 85 per cent.* In an extremely dry or cool season the tobacco does not pass through this curing process, but will be mottled with spots of green or yellow through the leaf. In such weather it is well to leave the barn open at night to let in the damp air, and close it in the day to keep out the dry air.

On the other hand, in a wet, warm season, there probably will be damage from pole-sweat or house-burn, which can be corrected to some extent by opening the barn on dry days and closing it on damp days and nights. If the tobacco becomes thoroughly limp under these methods then it is best to open the barn in any weather, as damp air circulating is not as bad as damp air stagnant in the barn.

In most sections a dirt floor is preferable, but where the country is flat and heavy fogs occur, a plank floor is to be recommended, especially where

---

* "The life activities of the tobacco plant practically cease at temperatures below 40° F., while they increase as the temperature rises until, at about 125° F., the living cells are rapidly killed. These activities are also greatly lessened by loss of water, and cease as soon as the leaf becomes dry. In practice, the most favorable temperatures for curing lie between the limits of 60° and 100° F., and the relative humidity should be about 85 per cent. Under these conditions the leaf will gradually lose its water, but will never be out of case or order, and the curing will proceed smoothly. If the humidity becomes much higher, pole-sweat will develop on the leaves most advanced in the curing, while if the humidity falls much below this figure the leaf will dry out too rapidly.

"In the second stage of the curing, when the leaf begins to turn brown, there is no longer any need for keeping the air in the barn so moist, and the relative humidity may be lowered to about 80 per cent., and later still further reduced to 65 or 70 per cent., until the stems are dry. So long as artificial heat is not used the temperature will never become too high for favorable curing, but it frequently becomes so low as to seriously interfere with the process." Bulletin No. 143, Bureau of Plant Industry, United States Department of Agriculture, Washington, D. C.: "Principles and Practical Methods of Curing Tobacco," by W. W. Garner.

early harvesting is practiced. When tobacco is cured early over a dirt floor there is danger of mould during the damp weather, after it is cured, before cool weather comes.

The common method of harvesting and curing tobacco in the sun-cured district is to cut and haul the tobacco to the barn immediately and hang it there to cure by air. But in some sections it is customary to scaffold in the field or at the barn for a few days. This is the better plan, as the sun kills the stalks and starts the curing quickly, which helps to make a sweet tobacco.

FIG. 6.—A good curing shed.

Formerly, when labor was plentiful and cheap, this was the universal custom in the sun-cured district, and in case of a sudden rain hands were available to put the tobacco quickly into the barn. In many sections this method of sun-curing is now impracticable on account of the high price and scarcity of labor.

When the labor can be had and the building so arranged as to permit, the best method is to scaffold the tobacco a few days in the open air, then hang in a shed open to the south to finish curing, after which it should be carried to a dry barn and hung until it is taken down to strip. The surest method is to build a shed with doors covering one entire end, then build trucks in sections that will just pass in and out of this shed. These trucks can be hung full of tobacco and rolled in and out of the shed on a track built for the purpose, according to the weather. When the truckful is cured it can be raised into the barn and refilled with green tobacco. It would not be practicable, of course, to cure the whole crop this way.

VISITORS ARE WELCOMED AT THE DISTRICT EXPERIMENT STATIONS; COME AND SEE THE RESULTS OF THE EXPERIMENTS. THE EXPERIMENT STATION IS ALSO GLAD TO ANSWER LETTERS FROM FARMERS ABOUT FARM PROBLEMS.

WHAT THE EXPERIMENT STATION CAN DO

Aside from reporting the results of its experiments, the Virginia Experiment Station offers to assist the farmers of the State, by correspondence, along the following lines:

1. The maintenance of soil fertility, including the rotation of crops and the use of manures and fertilizers.

2. The selection of varieties of farm crops—grasses, grains, fruits, vegetables—and the culture of these crops.

3. The identification of weeds, of varieties of fruits, and other plants.

4. The identification of the injurious insects and diseases of plants, and remedies therefor.

5. The breeding and care of farm animals, including the calculation of rations and the prevention and cure of diseases.

6. The planting and care of forest trees, and the management of the farm wood-lot.

7. The distribution of a very limited quantity of improved varieties of fruits, grains, and grasses, resulting from our breeding work. When any of these are ready, announcement will be made.

WHAT THE STATION CANNOT DO

1. It *cannot* analyze commercial fertilizers, seeds, limes, and soils. Such requests should be directed to the Commissioner of Agriculture, Richmond, Va.

2. It *cannot* report concerning outbreaks of contagious diseases of farm animals. Consult the State Veterinarian, Burkeville.

3. It *cannot* make an official inspection of orchards and nurseries for dangerous insects and diseases, this work being in charge of the State Crop Pest Commission. Address the State Entomologist, Blacksburg.

4. It *cannot* examine foods, drugs, feeding stuffs, and dairy products suspected of adulteration. Make request to the Dairy and Food Commissioner, Richmond.

5. It *cannot* analyze drinking water. Consult the State Department of Health, Richmond.

6. It *cannot* report concerning ores and minerals, marls and other deposits. Consult the State Geologist, Charlottesville, Va.

Address communications to

STATE EXPERIMENT STATION,
Blacksburg, Va.

AVAILABLE BULLETINS

VIRGINIA AGRICULTURAL EXPERIMENT STATION

Any or all of the following bulletins published by the Station will be sent free to anybody in Virginia who requests them, so long as the supply lasts. If you are interested in farming, have your name placed on our mailing list to receive new bulletins as issued. Bulletins not listed here are now out of print.

FARM CROPS AND FERTILIZERS

Bulletin 154—Inoculation and Cultivation of Alfalfa.
Bulletin 166—Improvement of Fire-Cured Tobacco.
Bulletin 168—Oats, Millets and Various Legumes.
Bulletin 174—Potato Growing.
Bulletin 175—Tobacco Investigations.
Bulletin 180—The Blue-grass of Southwest Virginia.
Bulletin 183—Work at the Tobacco Stations.
Bulletin 184—Impurities in Grass and Clover Seed Sold in Virginia.
Bulletin 187—Lime for Virginia Farms.
Bulletin 193—Grass Culture.
Bulletin 196—Crop Rotation and Fertilizer Experiments with Sun-Cured Tobacco.
Bulletin 197—Growing and Curing Sun-Cured Tobacco.
Bulletin 198—Crop Rotation and Fertilizer Experiments with Bright Tobacco.
Circular 1—Sugar Beets in Virginia.
Circular 3—Dates of Seeding Winter Grains.
Circular 4—Selecting Seed Corn.
Circular 5—Analyses of Sugar Beets in 1908.
Circular 6—Improving the Corn Crop.

LIVE STOCK AND DAIRYING

Bulletin 124—Sheep Scab.
Bulletin 125—Mange in Horses.
Bulletin 126—The Stomach Worm.
Bulletin 144—Stock and Poultry Powders.
Bulletin 156—Gluten and Cotton-Seed Meal, with Silage, Hay and Stover for Dairy Cows.
Bulletin 157—Silage, Hay and Stover in Beef-Making.
Bulletin 164—Stall Feeding Versus Grazing.
Bulletin 169—Protein Requirements for Dairy Cows.
Bulletin 170—Studies in Milk and Butter Production.
Bulletin 171—The Development of Grade and Cross-Bred Beef Cattle.
Bulletin 173—Finishing Beef Cattle.
Bulletin 176—Hog Feeding.
Bulletin 178—Causes of Loss of Lambs in 1908.
Bulletin 182—Silo Construction.
Bulletin 185—The Production of Clean and Sanitary Milk.
Bulletin 186—Tests of Hand Separators.
Bulletin 189—Some Diseases of Swine.
Bulletin 190—Coöperative Herd Testing.
Bulletin 194—Milk Standards.
Circular 8—The Dairy Cow and Her Record.

FRUITS AND VEGETABLES

Bulletin 130—Important Varieties of Apples.
Bulletin 132—Crab Apples.
Bulletin 142—The Bitter-rot of Apples.
Bulletin 143—The Composition of Apples.
Bulletin 146—Canning Fruits and Vegetables.
Bulletin 147—Bush Fruits.
Bulletin 155—Meteorological Data and Bloom Notes of Fruits.
Bulletin 177—Tomato Breeding and Varieties.
Bulletin 181—Wormy Apples.
Bulletin 191—Cabbage Club-root.
Bulletin 192—Tomato Blight and Rot.
Bulletin 195—Foliage Diseases of the Apple.
Circular 7—Fighting the Insect Pests and Diseases of Farm and Garden Crops.

ANNUAL REPORTS

These contain the results of the more technical investigations and are not sent to farmers except on special request. Reports for 1908, 1909 and 1910 now available.

Address correspondence to

AGRICULTURAL EXPERIMENT STATION,
Blacksburg, Virginia.

www.ingramcontent.com/pod-product-compliance
Lightning Source LLC
Chambersburg PA
CBHW062237220526
45471CB00009B/3524